The ROMAN FORT

Peter Connolly

Contents

Oxford University Press   1991

The Roman Fort

The last battle

The dull blast of a horn signalled the end of the fourth watch. The smouldering embers of a thousand camp fires, fanned by the cool morning breeze, glowed in the half-light. The night sentries relaxed at their posts waiting to be relieved. Soon soldiers were emerging from their tents yawning and shivering as they kicked over the embers and added wood to cook their breakfast.

The soldiers had been late to bed the previous night as they relived the fears and excitement of the day, glorifying their exploits and mourning the death of their comrades – for they had won a great victory.

The Roman governor, Agricola, had finally brought the Caledonians to bay and defeated them decisively at Mons Graupius in the summer of AD 83. Victory had come only after months of fruitless campaigning. Four regiments of Batavian soldiers and two regiments of Tungrians had led the charge. These soldiers were not Romans; they were auxiliaries, tribesmen from the frontier of the Roman empire who served with the Roman army. Wooden forts, manned by these auxiliary regiments, were built at strategic points after the battle to keep the newly conquered peoples under control.

This book is about one of these regiments, the First Tungrian Cohort (Cohors I Tungrorum), an eight hundred strong infantry unit, and the forts it occupied. The First Tungrian Cohort was larger than standard size and may have garrisoned the large fort at Fendoch near Perth covering the entrance to Glen Almond. Later the regiment occupied the fort at Vindolanda and finally Housesteads on Hadrian's Wall.

Legionaries building a barrack block and headquarters building (principia) in a timber fort. The buildings were made of timber and wattle, plastered over and painted to look like stone.

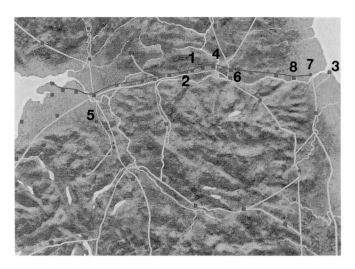

The north-west frontier

The Romans invaded Britain in AD 43. Only the south-eastern part of the country was occupied at first but the new province was gradually extended to include Wales and the north. The governor Agricola tried to complete the conquest of the island between AD 78 and 83, advancing up towards the Moray Firth where he won the battle of Mons Graupius.

While Agricola was conquering northern Britain the Romans had crossed the Rhine into southern Germany occupying the Taunus mountains and the Black forest. But in AD 85 the Dacians, living in the area now called Romania, crossed the Danube and defeated the Roman army there. Troops had to be withdrawn from Britain and Germany to meet the emergency and the advance in both countries was stopped. A chain of forts was established in the newly conquered German territory and the legions were pulled back beyond the Rhine.

A similar withdrawal was ordered in Scotland. Legion XX, which had probably been building a new base at Inchtuthil, north of Perth, moved back to Chester and Legion IX returned to York. Legion II Augusta occupied Caerleon in Wales while Legion II Adiutrix was withdrawn and posted to the Danube.

A new frontier zone, controlled by auxiliary regiments, was established in the Southern Uplands of Scotland centred on a large fort at Newstead with supporting units further south along the Roman road, known as the Stanegate, running between Carlisle and Corbridge. The First Tungrian Cohort withdrew to Vindolanda (Chesterholm) on the Stanegate. But it soon moved again.

Trajan, on becoming emperor in AD 98, began preparing for the conquest of the Dacians. More troops were withdrawn from Britain and the forts in the southern uplands of Scotland had to be abandoned. The First Tungrian Cohort returned to Vindolanda where a larger fort had to be built as a cohort of Batavians was already there.

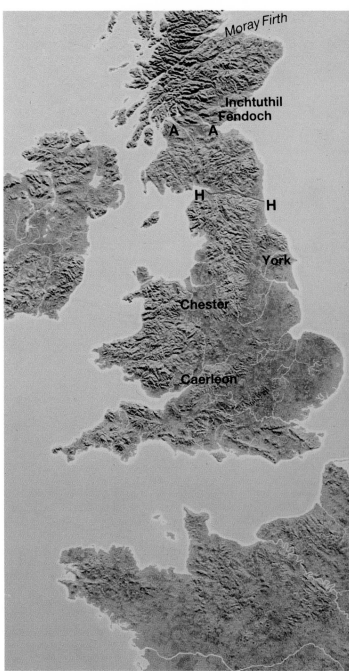

LEFT: map showing Hadrian's Wall (red line) with the forts and roads in the area.
1 Housesteads 2 Vindolanda
3 South Shields 4 Chesters
5 Carlisle 6 Corbridge
7 Newcastle 8 Benwell

BELOW: the north-west frontier of the Roman Empire (red line) in the second century AD. The frontier in Britain alternated between Hadrian's Wall (H-H) and the Antonine Wall (A-A). The frontier in southern Germany (below right) was just a chain of forts and watch towers. Hadrian moved the southern part of the line forward and erected a fence to control the Germans.

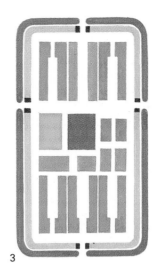

1

ditch
rampart
headquarters
barracks
commanding officer
granary
workshops etc.
gate

The three types of fort.
1 For about 200 men.
2 For about 500 men.
3 For about 800 men.

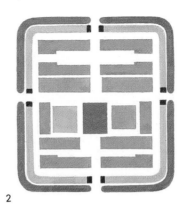

2

3

Frontier forts

The Roman word for a fort, castellum, means a fortified site where a small unit of soldiers were posted. Such forts were defended by a turf rampart or wall fronted by one or more ditches. During the great conquest period (210BC-AD107) forts varied greatly in both shape and size. After this forts were generally rectangular with rounded corners like a playing card and held between 200 and 800 men.

Inside the fort

The layout was almost always the same with two main streets dividing up the camp. The via praetoria led from the front gate to the headquarters (principia) and the via principalis joined the two side gates passing in front of the principia. The principia faced the junction of the two roads. The commanding officer's house (praetorium) was next to it. The soldiers' barracks were at the front (praetentura) and at the back (retentura) of the fort.

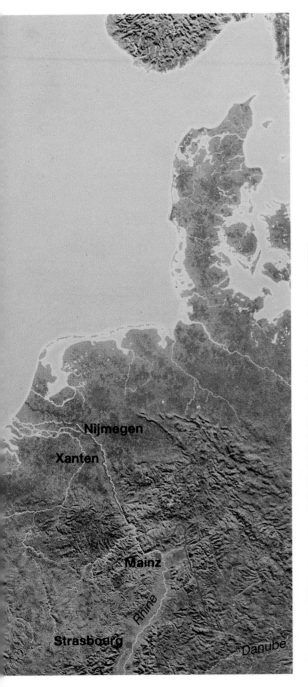

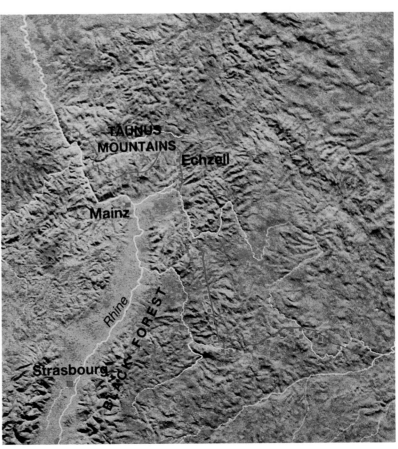

The Batavians and Tungrians at Vindolanda

The Batavians and Tungrians were Germanic tribes living near the mouth of the Rhine. Both supplied the Roman army with auxiliary troops. They had revolted against Rome in AD 69 and, when the revolt was put down, several Batavian and Tungrian cohorts appear to have been sent to Britain to get them out of the way. Later Agricola used these troops to spear-head his invasion of the north. Four Batavians and two Tungrian cohorts fought in the front line at the battle of Mons Graupius but the Roman historian Tacitus does not say which cohorts these were. The earliest evidence for the First Tungrian Cohort in Britain comes from a document found at Vindolanda which lists the number of men in the unit. This strength report, written on a wooden tablet, can be dated to about AD 90.

Having decided to regroup his forces along the Stanegate road the governor ordered the legions to build the necessary forts. Building was the job of the legions which possesed all kinds of craftsmen. The area was cleared and levelled, surveyors carefully marking out the two main roads and the various buildings – the headquarters, the commander's house, the barrack blocks, and the storehouses, workshops, stables etc.

Some legionaries dug the fortification ditches and stripped turf from the whole area to build the rampart. Others collected timber for the towers, gates, palisades and the interior buildings of the fort. They cut down the birch trees that grew plentifully in the area dragging them back to the fort where the carpenters set up their trestles and began cutting the timber to shape.

The defences complete, the legionaries began work on the interior buildings, constructing a skeleton of thick wooden posts filled in with wattle and daub. Finally each building was covered with thick plaster and painted to look like a stone building.

LEFT: the gateway of a timber fort. One cannot be sure about the reconstruction of such gateways for only the holes for the upright timbers are ever found.

BELOW: the fort at Vindolanda with the Stanegate running behind it in AD 105. The fort housed the First Tungrian cohort and a Batavian cohort. Only the line of the ramparts and the position of the house of one of the commanders (A) is known. Such forts had corner towers and interval towers. The timber to build these forts was gathered locally. Later Hadrian's Wall was built along the crest of the hills in the background.

ABOVE: timber and wattle walling discovered in the Roman fort at Valkenburg in the Netherlands. This type of wall could be prefabricated. The wattle was covered with rough plaster before the whole wall was surfaced with refined plaster and painted. Fragments of painted plaster suggest that the Romans painted such buildings to look as if they were made of stone.

Exciting discoveries

The excavation of the wooden fort at Vindolanda has brought to light a mass of every day things used in a Roman fort. The waterlogged site has preserved hundreds of leather and wooden items. Dozens of shoes for men, women and children have been discovered as well as pieces of tents, horse equipment and woollen garments. But the most exciting finds are over eight hundred wooden writing tablets giving invaluable information about life in the fort. These documents have given us the names of the units at the fort and their officers.

Hadrian's Wall

Trajan greatly extended the Roman empire, conquering Dacia (Romania), Armenia and Mesopotamia. The great conqueror died in AD 117 leaving the empire to Hadrian. Hadrian wanted to consolidate the empire. He stopped all further conquest and withdrew from Armenia and Mesopotamia. Now, for the first time, the Romans established fixed frontiers.

Hadrian visited many of the frontier provinces. In Britain he ordered the building of the great wall that bears his name. Built just north of the Stanegate, it stretched for 120km from the Solway Firth to the Tyne estuary.

The legions moved in dividing the area between them. The Second took the Vindolanda area. Traces of its work camps can be seen along the Stanegate, some being visible west of Vindolanda (see p.10).

The centurions appear to have moved into the fort itself where they could sleep in comfort while their men lived in tents. The wall was planned to be three metres wide and about five metres high. a fortlet was built every 1,500 metres with two turrets or watch towers in between. These fortlets, usually called mile castles, formed defended gateways through the wall.

Some legionaries dug the ditch, eight metres wide and three metres deep, in front of the wall whilst others quarried stone for the wall. The foundations of the whole wall were laid first. The fortlets and intervening turrets were completed next with a few metres of wall on either side of them. At this point there was a change of plan; the width was reduced to two and a half metres.

The cohorts manning the forts along the Stanegate took no part in building the wall but they might have helped dig the so called vallum. This was a ditch six metres wide and three metres deep excavated along the south side of the wall to define the military zone. The vallum is usually only a short distance back from the wall leaving just enough space for the marching road which joined all the fortlets and turrets.

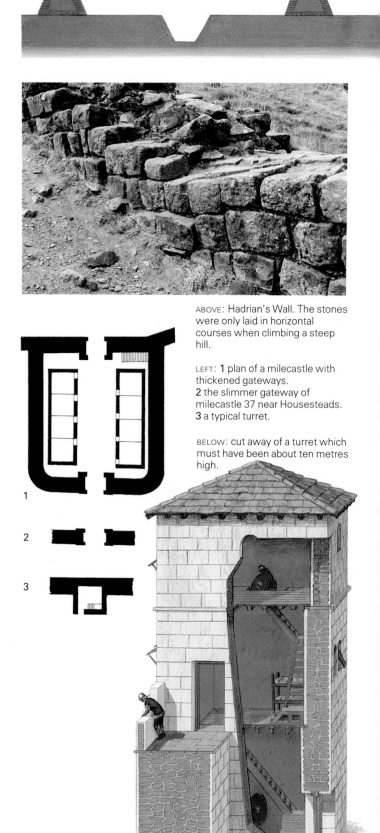

ABOVE: Hadrian's Wall. The stones were only laid in horizontal courses when climbing a steep hill.

LEFT: **1** plan of a milecastle with thickened gateways.
2 the slimmer gateway of milecastle 37 near Housesteads.
3 a typical turret.

BELOW: cut away of a turret which must have been about ten metres high.

1

2

3

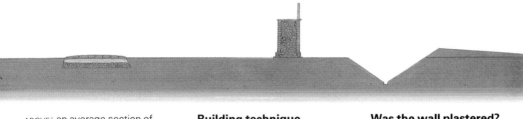

ABOVE: an average section of Hadrian's Wall with its fronting ditch, vallum and marching road. The broad gap between the ditch and the wall is to prevent the wall collapsing into the ditch. The earth from this ditch was dumped beyond it and levelled off to increase the depth of the ditch.

BELOW: milecastle 37 just west of Housesteads.

Building technique

Hadrian's Wall has outer facings of rough cut stone with a core of rubble and mortar. The stone courses generally run parallel to the ground but the stones had to be laid in horizontal courses when the wall climbed a very steep incline. Only the gateways in the forts and milecastles are made of carefully cut masonry.

Was the wall plastered?

The Romans plastered everything, leaving only the finely cut masonry uncovered. The wall was probably plastered and painted with red lines to look like square cut masonry. Such painted plaster has been discovered in Germany. The plaster would only have covered the rough cut stone leaving the gateways clear.

Milecastles

The milecastles, built at regular intervals of one Roman mile (1500 metres), served as fortified crossing points. Some milecastles have gateways that are considerably thicker than their adjoining walls (see opposite), implying that they supported a tower above the gate. Others, such as milecastle 37 just west of Housesteads, have much slimmer gates which could only have supported a very narrow tower. The choice seems to have depended on which legion built it; the Second Legion seems to have preferred the narrow type.

The auxiliaries move up

At some point after the foundations of the wall had been laid there was another change of plan. Probably a new governor had decided to 'do it his way'. The governor decided the forts along the Stanegate were too far away to service the wall and he ordered new forts to be built along the wall. One of these new forts was at Housesteads, three kilometres north-east of Vindolanda. We know that this move was a change of plan because a turret and a short stretch of wall had to be demolished to make room for the new fort. The foundations of the wall turret are clearly visible within the northern part of the fort.

We do not know which cohort garrisoned the fort at Housesteads first but the fort was built to house a 800-strong unit. The First Tungrian cohort was such a unit and it certainly garrisoned Housesteads in the third century. For the purposes of this book it will be assumed that the unit moved straight from Vindolanda to Housesteads.

The fort at Housesteads was built on a steep, windswept slope. The main source of water for the fort was a small stream, the Knag Burn, running down a valley a hundred metres east of the fort. One of the fatigues of the soldiers must have been carrying innumerable buckets of water up the steep slope to replenish the cisterns in the fort. The barrack blocks, arranged lengthways along the slope, were built on stone foundations possibly with the upper part made of timber, wattle and daub. The forts like the wall were built by the legionaries. One of the internal buildings at Housesteads, the commander's house (praetorium), was erected on a very steep slope and took considerable engineering skill to build.

Most of the camp followers would have moved up to the wall with the cohort and set up their village (vicus) outside the walls but some appear to have stayed behind in the more pleasant climate at Vindolanda which continued to be used periodically by different units.

The Housesteads/Vindolanda area.
1 Housesteads fort.
2 Housesteads parade ground.
3 Tribunal. 4 Bath house.
5 Knag Burn.

6,7,8 Milecastles.
6 Milecastle 37.

9 Military road.
10 Vindolanda - Housesteads road.
11 Vallum. 12 Vindolanda.
13 Vindolanda parade ground.
14, 15, 16 Large Roman camps.
17, 18 Small practice camps.
19 Marching camp.
20 Stanegate.
21 Signal station, probably used to relay messages from the mile castles and turrets to Housesteads and Vindolanda.

The basic unit of the Roman army was the century, 80 men commanded by a centurion. A century had its own standard (signum) carried by a standard bearer (signifier). It also had a second in command (optio) and guard commander (tesserarius).

A normal cohort (cohors quingenaria) consisted of six centuries. Some cohorts, such as the First Tungrian, were increased to ten centuries (cohors milliaria). Each cohort had its own commander, a prefect, and its own flag (vexillum).

ABOVE: remains of a vexillum from Egypt.

BELOW: a century, 80 men commanded by a centurion.

The soldiers

The Roman army was made up of two types of soldier, legionaries and auxilaries. Legionaries were Roman citizens. Auxiliaries were frontier people, non-Romans, who served with the Roman army. Originally the First Tungrian Cohort had been composed of Tungrian tribesmen whose native fighting skills were useful to the Romans. In time they adopted a more regular method of fighting and new recruits could be drawn from the local British population.

A man wishing to join a cohort needed a reference from someone with influence. A reference, written on a wooden tablet, was found at Vindolanda. It reads: 'He is my dear friend and a capable person. He has requested me to recommend him to you . . .'

The new recruit had to be a free man, medically fit and, before being accepted, he had to undergo rigorous basic training. He was then assigned to a century and began his 25 years service.

The new soldier had to do all the dirty jobs but in a few years he could become an immunis with a special job such as a medical orderly, clerk, trumpeter, etc. He received no more pay but his name was removed from the fatigues list. Further promotion gained extra money and the rank of principalis. The century's guard commander and standard bearer were on pay and a half. The second in command of a century was on double pay.

Auxiliaries received only one third the pay of a legionary but they seem to have been reasonably well off. They had to pay for their clothing and equipment. They generally wore a tunic and calf length trousers. A letter found at Vindolanda mentions also underpants and socks which would have been essential winter clothing on the northwest frontier.

Soldiers were forbidden to marry but many had common law wives and children living in the village. On retirement an auxiliary soldier was normally given Roman citizenship. This included any children he had or would have in the future.

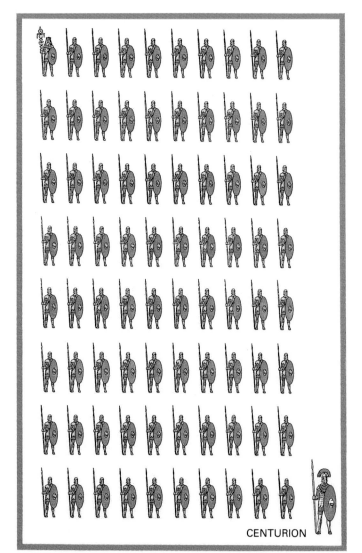

CENTURION

COHORS QUINGENARIA

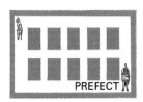

COHORS MILLIARIA

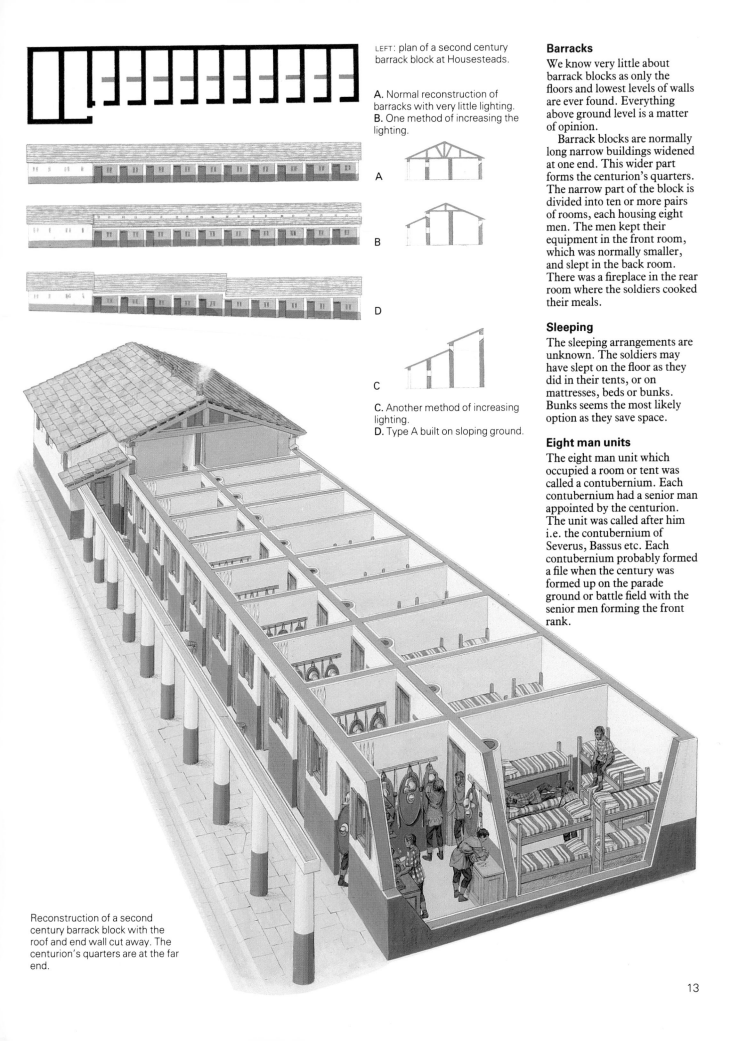

LEFT: plan of a second century barrack block at Housesteads.

A. Normal reconstruction of barracks with very little lighting.
B. One method of increasing the lighting.

A

B

D

C

C. Another method of increasing lighting.
D. Type A built on sloping ground.

Barracks

We know very little about barrack blocks as only the floors and lowest levels of walls are ever found. Everything above ground level is a matter of opinion.

Barrack blocks are normally long narrow buildings widened at one end. This wider part forms the centurion's quarters. The narrow part of the block is divided into ten or more pairs of rooms, each housing eight men. The men kept their equipment in the front room, which was normally smaller, and slept in the back room. There was a fireplace in the rear room where the soldiers cooked their meals.

Sleeping

The sleeping arrangements are unknown. The soldiers may have slept on the floor as they did in their tents, or on mattresses, beds or bunks. Bunks seems the most likely option as they save space.

Eight man units

The eight man unit which occupied a room or tent was called a contubernium. Each contubernium had a senior man appointed by the centurion. The unit was called after him i.e. the contubernium of Severus, Bassus etc. Each contubernium probably formed a file when the century was formed up on the parade ground or battle field with the senior men forming the front rank.

Reconstruction of a second century barrack block with the roof and end wall cut away. The centurion's quarters are at the far end.

The commanding officer

We know the names of several of the commanders at Vindolanda. One, Flavius Cerialis, was in command of the First Tungrian Cohort about AD 105.

Cerialis came from a wealthy family living in some Romanised part of the empire such as the Roman colony at Arles in the south of France. He must have been a senior magistrate (duumvir) of his town as a first step in his career. He then had to find an influential patron who could recommend him to the governor of a province.

Accepted by the governor, he was probably made commander (prefect) of a 500 strong auxiliary infantry cohort. Although about 30 years old, he was unlikely to have seen military service before and had to rely heavily on the advice of his centurions and any friends with military experience he might have brought with him. Cerialis obviously coped well for after about three years he was promoted to commander of the larger 800 strong unit, the First Tungrian Cohort.

A cohort commander usually brought his wife and family with him. They lived in lavish quarters occupying nearly 10% of the built up area of the fort. Their house was the commander's official residence (praetorium) and had all the amenities befitting his status, heated rooms, bathing suite and plenty of servants' accommodation. But it was a lonely life; they could not mix socially even with centurions in the fort for they were members of the upper class (equites). Their only social life was with the commanders of the other forts in the area. This is dramatically brought to life by a letter written to Cerialus' wife, Sulpicia Lepidina, by her friend, Claudia Severa, the wife of another fort commandant. Severa begs Lepidina to come to her birthday party on the 12th September. The letter ends with greetings to Cerialis from Severa's husband, Aelius, and their little son.

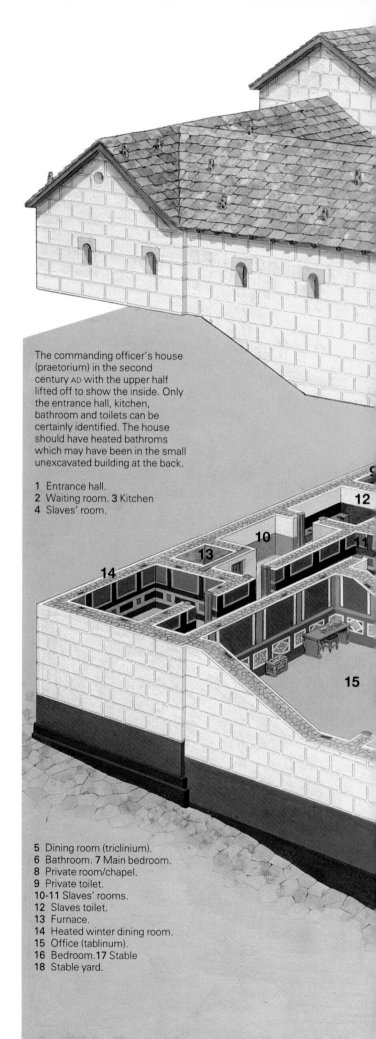

The commanding officer's house (praetorium) in the second century AD with the upper half lifted off to show the inside. Only the entrance hall, kitchen, bathroom and toilets can be certainly identified. The house should have heated bathroms which may have been in the small unexcavated building at the back.

1 Entrance hall.
2 Waiting room. 3 Kitchen
4 Slaves' room.

5 Dining room (triclinium).
6 Bathroom. 7 Main bedroom.
8 Private room/chapel.
9 Private toilet.
10-11 Slaves' rooms.
12 Slaves toilet.
13 Furnace.
14 Heated winter dining room.
15 Office (tablinum).
16 Bedroom. 17 Stable
18 Stable yard.

The principia

The headquarters (principia) stood in the centre of the fort at the junction of the two main streets, the via principalis and via praetoria. The front half of the principia consisted of a courtyard surrounded by a verandah. The official notices of the cohort were probably posted on the walls of the courtyard and soldiers would have gathered to read them in the shade of the verandah.

The cross-hall

The cross-hall (basilica), was a high draughty building open to the weather on the courtyard side. The soldiers gathered here to be addressed by the prefect. The raised platform (tribunal) from which the commander spoke was at the north end of the hall. It was mounted by a flight of steps on the right.

The headquarters (principia)

The prefect rose early so that he could be at headquarters by sunrise for his daily meeting with his centurions. Here he was assisted by his adjutant (cornicularius) with his staff of junior officers (beneficiarii) and clerks (librarii). He was really only required to give overall directions, hear and grant or refuse the requests of his men and give judgement on matters of discipline. The ten centurions and the adjutant ran the fort.

The headquarters building (principia) was divided into three main parts. The front half was an open courtyard surrounded by a verandah. Beyond this was a high roofed hall, usually called the cross hall, with a raised dais (tribunal) from which the commander would address his assembled troops.

At the far side of the cross hall was a range of five rooms. The central room was the chapel (aedes) where the standards and a statue of the emperor were kept. The regimental savings were also kept here, usually in a strong box under the floor or in an underground chamber beneath the chapel. The chapel was the most sacred room in the fort and was permanently guarded.

The two adjoining rooms to the right of the chapel were probably used by the adjutant and his clerks and the room to the left by the standard bearers (signiferi). They acted as bankers for their centuries. The other room may have been the armoury. The three central rooms rooms, including the chapel, were separated from the hall by a low stone screen topped by a metal grille which allowed a free view into the chapel and permitted the adjutant and standard bearers to carry on their duties without letting the soldiers into their rooms. The remains of these stone screens from Vindolanda are worn deep between the bars where the soldiers had rubbed the stone when receiving or handing in money and docments through the grille.

The principia at Housesteads in the second century AD. The centurions met the prefect here each morning to report on the state of their centuries and to receive their orders for the day. The chapel, in the centre at the back, was the focal point of the fort. The standards and the image of the emperor were kept here.

The chapel

The chapel (aedes or sacellum) was the most important room in the fort. It stood directly opposite the entrance to the cross-hall so that any soldier passing the entrance of the principia could see it. The standards and the bust of the emperor were kept there. It was kept permanently guarded by a centurion and six soldiers, two soldiers probably standing guard for two hours in every six.

Administration

The adjutant (cornicularis) probably occupied the pair of rooms to the right of the chapel. All the fort documents were kept in the adjutant's office. The adjoining room was probably used as a store room. The Roman army, like most armies, kept written records of absolutely everything. Thousands of these documents, written on papyrus, have been found buried in the sands of Syria and Egypt. Hoards of similar documents, but written on wooden tablets, have been discovered at Vindolanda.

Banking

The standard bearers were the fort bankers. They were responsible not only for the military chest but also the individual soldier's savings. All soldiers were compelled to save part of their money for which they received a receipt. The standard bearers also kept an account book listing individual soldiers and how much pay they had received after deductions for clothing etc. The money was kept in an iron bound wooden chest beneath the floor of the chapel. Some forts, such as Chesters, nine kilometres east of Housesteads, had an underground strong room.

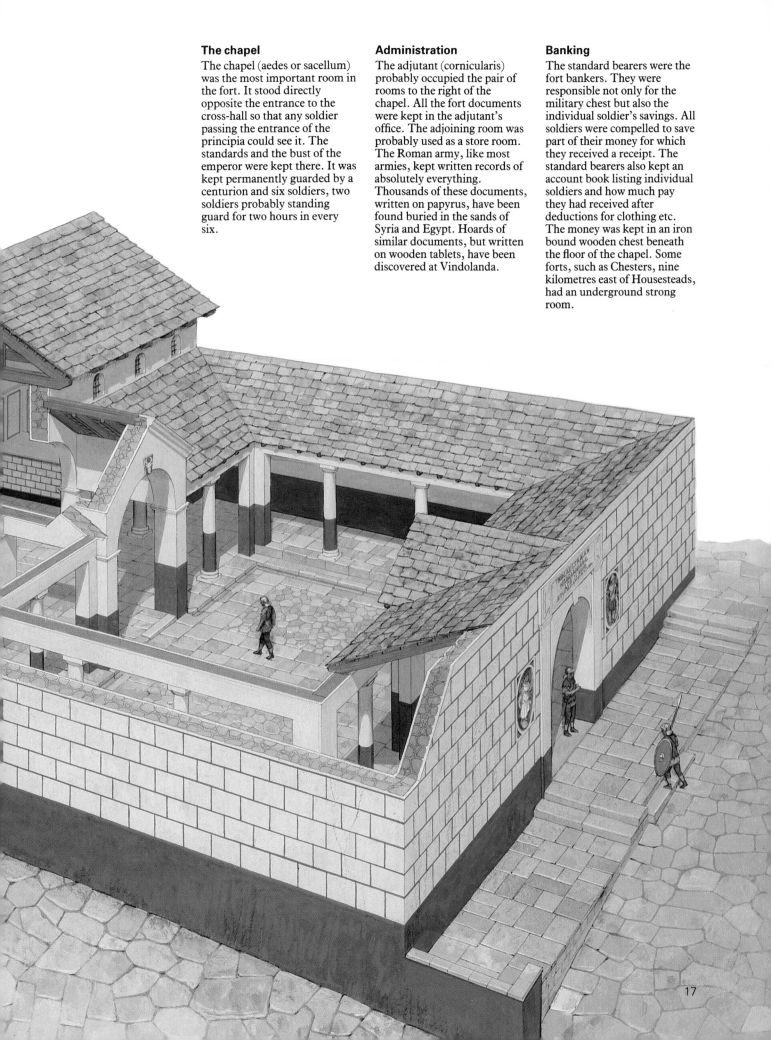

The daily report

Each morning the centurions reported the state of their centuries to the prefect in writing. Several of these reports have been found in Syria. They record the state of the Twentieth Palmyran Cohort stationed at Dura Europos in the third century AD. They list the numbers available for duty and why others are not available: i.e. 'sent to collect wheat', 'collecting wood for the bath house' and so on. They also give the password for the day, the work allocation and the names of the men who will stand guard at the chapel of the standards.

The hospital

There is a large courtyard building behind the principia which is generally thought to be a hospital. The reason for this identification is mainly the discovery of the tombstone of Anicius Ingenuus, a medical officer stationed at Housesteads in the third century AD. The building could equally well have been a workshop.

BELOW: the centurions meeting the prefect to deliver their daily status report. In Britain these reports were written with a stylus on wooden tablets covered with wax.

Morning report

The centurions gathered in the cross hall of the headquarters at day break. They handed in lists of the men available for duty with details of those who were absent or ill. The adjutant's clerk totted up the lists and entered them in the day book, listing the junior officers by their rate of pay.

The senior centurion called his colleagues to attention as the prefect entered. The adjutant picked up the wooden tablet on which the clerk had listed the men available for duty and read it out:

"March 24th. 484 infantrymen, including nine centurions, seven double-pay and two pay-and-a-half officers present on the base of the First Tungrian Cohort."

The prefect gave the password for the day:

"Holy Mercury, chosen from the seven planets."

The clerk noted this together with the prefect's full name. The prefect now turned his attention to the list, asking about those who were away and the new arrivals at the fort. The senior centurion explained that one of his colleagues was still away with his century at York, 260 men were on sentry duty at the mile castles and turrets and 35 were in sick bay. The clerk listed these points.

The senior centurion discussed the work for the day with the prefect, pointing out that supplies were running low and that help was needed in the hospital. The work was split up amongst the centurions. Each took turn· to mount guard at the chapel of the standards. The work allocated, the centurions swore to do their duty:

"We will do whatever may be ordered and be ready at every command."

The clerk entered this with a list of the men, a standard bearer (signifer), trumpeter (tubicen), priest (sacer), guard inspector (tesserarius) and two ordinary soldiers (milites), who were to stand watch with the centurion at the standards of 'Our Lord the emperor, Publius Aelius Hadrianus.'

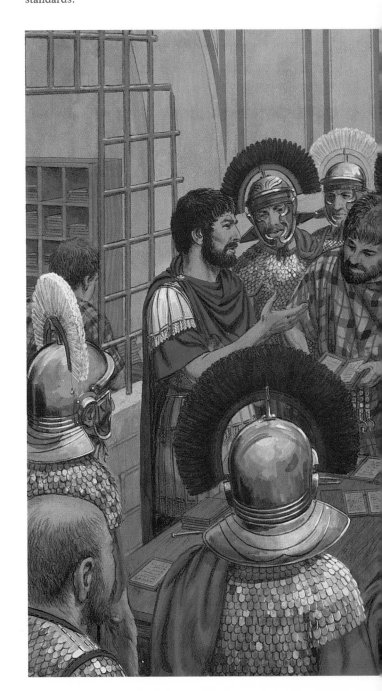

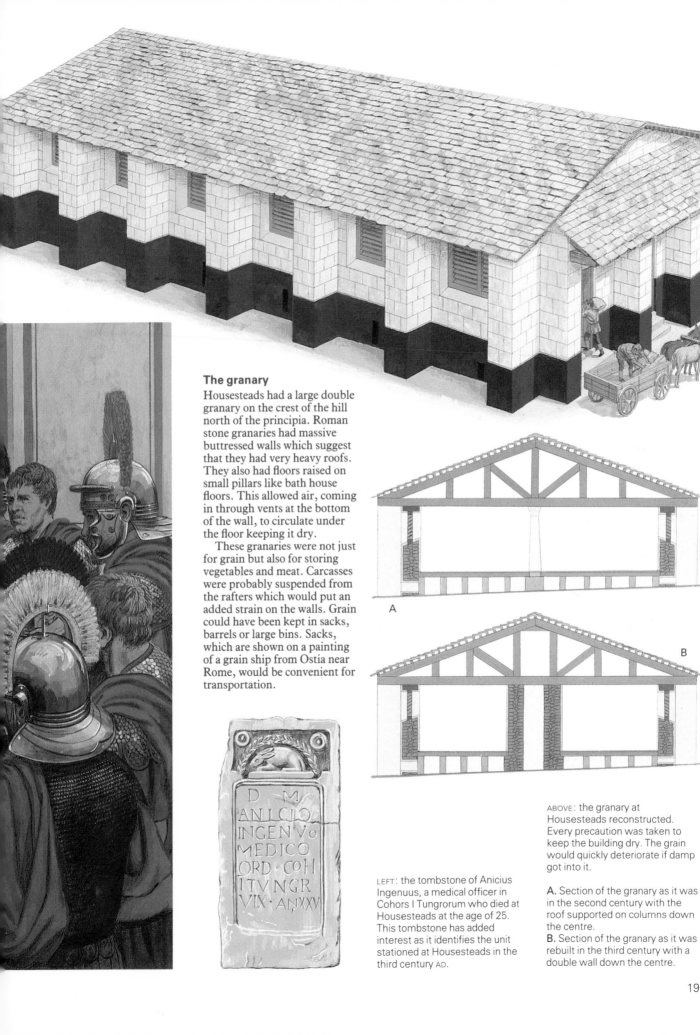

The granary

Housesteads had a large double granary on the crest of the hill north of the principia. Roman stone granaries had massive buttressed walls which suggest that they had very heavy roofs. They also had floors raised on small pillars like bath house floors. This allowed air, coming in through vents at the bottom of the wall, to circulate under the floor keeping it dry.

These granaries were not just for grain but also for storing vegetables and meat. Carcasses were probably suspended from the rafters which would put an added strain on the walls. Grain could have been kept in sacks, barrels or large bins. Sacks, which are shown on a painting of a grain ship from Ostia near Rome, would be convenient for transportation.

A

B

LEFT: the tombstone of Anicius Ingenuus, a medical officer in Cohors I Tungrorum who died at Housesteads at the age of 25. This tombstone has added interest as it identifies the unit stationed at Housesteads in the third century AD.

ABOVE: the granary at Housesteads reconstructed. Every precaution was taken to keep the building dry. The grain would quickly deteriorate if damp got into it.

A. Section of the granary as it was in the second century with the roof supported on columns down the centre.
B. Section of the granary as it was rebuilt in the third century with a double wall down the centre.

19

The centurions
The First Tungrian Cohort had ten centurions. These men, who had risen from the ranks, were selected for their bravery, reliability and education. The senior centurion (ordo princeps) was the top professional soldier in the fort. The prefect relied heavily on his advice.

BELOW: the chief centurion watching his troops as they march out for a ceremonial parade in front of the prefect. The column is led by the standard bearers and musicians.

Easy living
A lavishly decorated dining room (right) belonging to an auxiliary centurion was disovered at Echzell, one of the frontier forts in southern Germany. This illustrates the high standard of living centurions enjoyed.

RIGHT: the centurion's dining room at Echzell.

All in a day's work

The centurions called out those available for general duties. Relief parties were selected to take over at the six mile castles and their turrets along the wall and guards were chosen for the four gates of the fort. One century was ordered down to the parade ground for weapons training. Another two centuries were sent out on their thrice monthly route march. They would entrench a small camp for the night and march back the following day.

An escort was selected for the prefect who was going away for a few days. A working party with three wagons under the command of an optio was dispatched to the supply base at South Shields to bring up supplies. Four men were ordered to report to the hospital as stretcher bearers and fatigue parties were selected for the dirty jobs: twelve men to sweep out the headquarters, fifteen to clean the bath house and three to clear a blockage at the latrines.

These tasks were entered on the duty roster by the clerks of the centuries. The soldiers' names were listed in a column down the left side and their duties filled in day by day across the sheet.

The chief centurion (ordo princeps), who took command in the absence of the prefect, checked the men had reported for their various duties and were doing them properly. He visited the hospital and the latrines and then strolled out through the south gate, inspecting the sentries as he went. He walked down the hill, through the civil settlement to the parade ground where the soldiers were training with wooden swords and spears against the stakes. He mounted the tribunal, flanked on either side by altars, stood there watching for a while and then returned to the fort.

The centurions were promoted from the ranks. Every soldier dreamed of becoming a centurion. They were well paid and lived in comparative luxury in a suite of rooms at the end of their barracks.

The parade ground

Every Roman fort had a parade ground, a large flat area where soldiers could train and parade for the inspection of the prefect.

On one side was a raised platform (tribunal) from which the prefect reviewed his troops. Altars and shrines dedicated to Jupiter, Mars and Victory flanked the tribunal.

Parade grounds were usually surfaced with shingle and might have had permanent training stakes set up on it. These were sturdy timbers, the height of a man, against which the soldiers practiced with their weapons.

Housesteads parade ground

The site of the parade ground at Housesteads has never been identified. This is rather strange as there is really only one possibility, a large flat area at the southern foot of the hill on which the fort stands. There is a stony, flat-topped mound about five metres square and two metres high at the bottom of the slope. This could well be the remains of the tribunal. The hill on the south side appears to have been cut away to square off the site. The whole area is marshy today but it would certainly have been drained in Roman times.

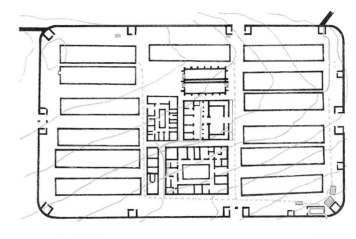

ABOVE: plan of the fort at Housesteads showing how the waste water was channelled down to the latrine. Known drains are shown in solid blue.

BELOW: plan of the Housesteads latrine.
A and B The large tanks.
C The small tank.

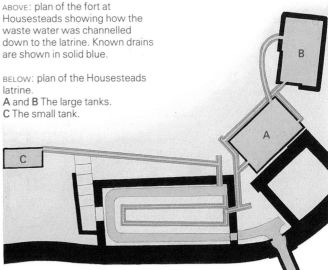

Water and waste

Housesteads fort was built on the crest of a ridge and seems to have had no natural water supply. About 3,000 litres of water were required every day for drinking, cooking and watering the animals. There may once have been a small spring in the north west of the fort for water still seeps through the rock but most of the water had to be obtained from wells or dragged up the steep slope from the Knag Burn 100 metres east of the fort. Rain water, dripping off the eaves, was also collected by channelling it into cisterns.

Waste and surplus rain water flowed away down the hillside through drains under the roads. One of these ran under the hospital and alongside the commander's house where it was used to wash out the toilets. The drains finally converged at the south-east corner of the fort, where they flushed out the public toilets.

The toilets were the normal Roman type, a large room with a deep sloping stone-lined trench about a metre wide running along the inside of three of the walls. The water flowed into the trench at the north-east corner of the building, ran along the north side, round the western end, along the south side before passing out through a hole in the fort wall.

The toilet seats, which rested on stone supports set into the wall above the trench, were like long benches made of stone or wood and boxed in at the front. The benches were pierced with a series of key-hole shaped openings which extended down the front panel. This front opening was used for cleaning oneself with a sponge on a stick – the Romans did not use toilet paper. In front of the seats was a shallow trough, permanently filled with running water in which the sponge sticks were washed.

The Housesteads' toilet could seat about thirty people at once. There were no partitions separating the soldiers. Such buildings were normally open to the sky with a narrow sloping roof over the benches to protect the users from the rain.

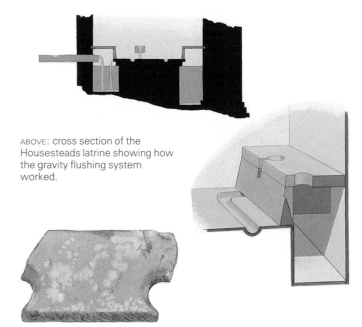

ABOVE: cross section of the Housesteads latrine showing how the gravity flushing system worked.

ABOVE: part of a toilet seat from South Shields at the eastern end of Hadrian's Wall.

ABOVE: a section of one of the public latrines at Vaison-la-Romaine in southern France. This is a typical Roman latrine with stone seats allowing two Roman feet per person.

ABOVE: a stone lined drainage channel running beneath the street just outside the praetorium at Housesteads.

ABOVE: a sponge stick.

BELOW: a drain cover from the courtyard of the principia at Housesteads.

Flushing out

Surplus rain water dripping from the eaves of the buildings at the east end of the fort at Housesteads was channelled into two large cisterns just ouside the latrine. This was used to keep a permanent flow of water in the shallow channel used for washing the sponge sticks. This was operated by gravity. The water flowed in at the east end, ran along the south side, round the west end and back along the north side where it was channelled under the bench into the main trench. Here it flowed back along the north side, around the west end and along the south side before passing out through a hole in the fort wall.

Surplus and waste water from the rest of the fort was collected in a tank just west of the latrine. This water was used to flush out the main trench at regular intervals. There would have been no shortage of water for flushing out in the winter but in the summer most of the water would have to be carried up from the Knag Burn.

BELOW: a reconstruction of the latrine at Housesteads as it was in the third century AD. It has been cut away to show the trench underneath.

Roman toilet seats

Roman public toilets have been excavated all over the Roman world. They all work on the same basic principles as the Housesteads example. The seating from the Housesteads latrine is unfortunately missing but there seems to have been a standard allowance of two Roman feet (60cm) per person. The three public latrines at the Roman town of Vaison-la-Romaine in southern France allow exactly two Roman feet. The fragmentary stone seat from South Shields in northern Britain also allows this amount. The public toilets at Ostia, the port of Rome, allow space for large and small people, the space between the holes varying from 49 to 70cm.

Roofing

Roman public latrines were probably open to the sky with a narrow roof above the seats. The upper part of these buildings is always missing and only when the roof was supported on columns, as it was in Athens and Rome, can one be sure of the existence of a roof. A narrow sloping roof over the seats did not require columns to support it and it is reasonable to assume that most if not all toilets had such roofs. This would be especially necessary in a rainy climate such as Britain's.

Another emperor: another policy

Hadrian died in AD 138 leaving the empire to Antoninus Pius who decided to reoccupy southern Scotland and construct a new wall across the narrow 50km wide isthmus between the Clyde and the Forth.

The new wall was built of turf and fronted with a ditch twelve metres wide. Closely spaced timber forts, seldom more than three kilometres apart, were built along the short frontier making a break-through almost impossible. The legions were left far behind. The nearest, the Sixth at York, was more than 300km, about ten day's march, away. Hadrian's Wall was now obsolete; causeways were built across the vallum and the gates were removed from the mile castles allowing free travel.

Successive emperors continued to change their policy. Some twenty years later the Antonine Wall was abandoned and the troops reoccupied Hadrian's Wall but within a few years they were back on the Antonine Wall again. By the beginning of the third century the army appears to have been back on Hadrian's Wall but almost immediately moved north again as the emperor Septimius Severus made one last attempt to conquer Scotland. Severus died in AD 211 before his work was completed. The army retreated to Hadrian's Wall never to invade Scotland again.

The First Tungrian Cohort marched south again and occupied the fort at Housesteads. For the first time we know for certain it was there as three inscriptions found at the fort mention the unit by name. The fort was rebuilt. Most of the remains now visible are third century. The verandas of the headquarters were blocked in completely enclosing the cross hall and forming a series of rooms along the north, west and south sides of the courtyard.

Vindolanda, which had seen many units come and go, was now occupied by the Fourth Gallic Cohort. Excavations outside the fort have revealed a whole village which tells us much about how the soldiers spent their spare time.

LEFT: reconstruction of the west gate, cut away to show the inside. There was a guard chamber on either side with a tower above. The west gate was probably the gate most used by heavy traffic as it leads to the granaries. Deep ruts in the threshold of the south gate show that it was used by wheeled traffic in spite of the very steep incline. The gates, which swung on pivots, were closed and barred from the inside.

Shops

Some of the long narrow buildings in the Housesteads vicus have one end open to the street. These are shops or bars. One has a long stone threshold with a slot along it. This type of shop front is well known from Pompeii in Italy. The front of the building was closed with wooden slats fitted into the slot in the threshold (see right). When open, merchandise was hung up above the counter which blocked most of the entrance.

RIGHT: **A** Roman shop closed with wooden slats.
B The threshold and counter.
C Wooden slats.

In the village

The camp followers began to arrive in their carts a few days after the soldiers. These were the merchants, shop keepers, craftsmen, girlfriends, wives, children and all the others who depended on the army for their existence. They followed the cohort whereever it went, setting up their stalls and tents on either side of the roads outside the fort. In time the tents were replaced by permanent buildings and the vicus was born.

Soldiers spent most of their spare time in the vicus for here were the bars, restaurants, shops and places of entertainment that could not be found inside the fort.

For hundreds of years soldiers had been forbidden to marry but most had girlfriends living outside the walls. Many set up a home in the vicus, marrying their girlfriends by the local Celtic rights and raising families. These marriages were illegal but nobody tried to stop them. Auxiliary soldiers were normally given Roman citizenship on retirement. Being no longer soldiers, they were able to marry their common law wives and adopt their children giving them citizenship too.

But now all this changed. Early in the third century soldiers' marriages were made legal and citizenship was granted to all free men within the empire. Most soldiers now had wives and families in the vicus and a soldier's son was expected to follow him into the army.

The vicus at Housesteads appears to have been built around the south, east and west gates but only the southern area has been excavated. The buildings are mainly long narrow structures often with a shop front facing the road. Archaeologists excavating the vicus in 1932 found two bodies buried under the floor of one of these buildings. They had clearly been murdered. The buildings at Vindolanda are more interesting as they include an inn and a bath house besides the normal shops and houses.

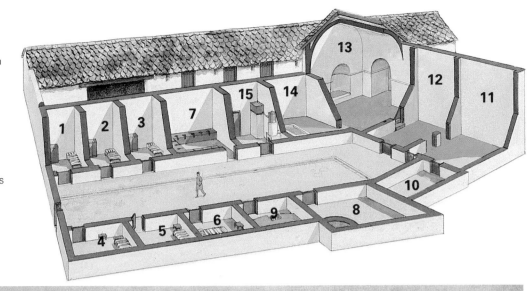

RIGHT: the inn (mansio) at Vindolanda had heated bathrooms, dining room and a communal toilet.

The inn was used by travellers on the Stanegate, the main road across northern Britain.

1-6 Guest rooms.
7 Communal toilet. 8 Kitchen.
9 Dining room.
10-15 Bathing rooms.
10 Lobby. 11 Changing room.
12 Cold room. 13 Warm room.
14 Hot room. 15 Boiler room.

The stables and servants quarters are at the back.

BELOW: Vindolanda about AD 200. The fort is in the background.
1 The inn. 2 The bath house.
3 The Stanegate.

ABOVE: the hypocaust. The floor was supported on pillars so that the hot air could circulate below it before being drawn up the flues in the wall.

A section of a clay flue pipe used to draw the hot air up the walls.

The bath house at Vindolanda

The bath house at Vindolanda was fully excavated in 1970-71. It is one of the few buildings so far discovered in Britain which still had plaster on the outside. It is a typical later Roman bath house having hot steam rooms and a hot dry room.

The Vindolanda bath house and the one at Chesters, fourteen kilometres to the east, are the only two bath houses along Hadrian's wall to have been fully excavated. Features of the Chesters bath house are shown here when the evidence at Vindolanda is missing.

The heated rooms

The floors of the heated rooms, which were made of flag stones resting on small stone pillars, were surfaced with a mixture of brick chips and cement. There were rectangular slots, about 35cm wide and 12cm deep, running up the walls of the heated rooms. These held the clay pipes which drew the hot air up to the chimneys in the roof. The same system was used at the inn and the commander's house at Housesteads. Heated rooms had vaulted ceilings. Those at Chesters were hollow which helped to insulate the rooms in the same way as double glazing.

The Vindolanda bath house had a stone tiled roof. Many pieces of these large grey sandstone tiles were found during excavation.

The soldiers of Legion VI Victrix built the main part of the building, the hot, warm and cold rooms. They probably left the rest, the changing room and the toilet, to be built by the auxiliaries in the fort. Certainly the two parts of the building were constructed separately.

Relaxation for men and women

The baths were not just for bathing but for general relaxation. Snacks and drinks were served there and the soldiers gambled with dice and knuckle bones. Many hair pins, beads and combs were found during the excavations which suggest that women also used the baths, probably at special times when the men were busy in the fort.

LEFT: the cubby holes for clothes in the changing room of the baths at Chesters.

Somewhere to relax: the baths

Most forts had a bath house outside the walls. There is one, which has never been excavated, on the far bank of the Knag Burn at Housesteads.

The baths at Vindolanda were built by a working party from the Sixth Legion about AD 150. The auxiliaries had gained some building skills by this date but the engineering problems of a bath house with its underfloor heating and flues up the walls was beyond their capabilities.

Bath houses consisted of cold, warm and hot rooms. The hot and warm rooms had raised floors supported on small columns about a metre high. The furnace was situated next to the hot room, the heat being drawn through a hole in the wall into the area beneath the floor of the hot room. It then passed through another hole in the far wall into the space under the floor of the warm room. The heat was drawn up the walls through pipes, the hot air escaping through chimneys in the roof. The same furnace heated water for the hot and warm plunge baths.

A bather entering the changing room (apodyterium), stripped off, leaving his clothes in a cubbyhole in the wall. He then entered the cold room (frigidarium) where he could take a cold bath. He went on either to the very hot dry room (laconicum) or into the warm steam room (tepidarium). This adjoined the hot steam room (caldarium) where he could take a hot bath.

The hot clammy atmosphere of the caldarium made the bather perspire in the same way as a modern Turkish bath. A slave scraped the bather with a strigil to remove ingrained dirt from his skin. After being scraped clean the bather might have a massage and a cold bath. The soldiers spent much of their spare time at the baths for it was more than just a bathing establishment. It was more like a club where they could meet their friends and relax, away from the discipline of the fort.

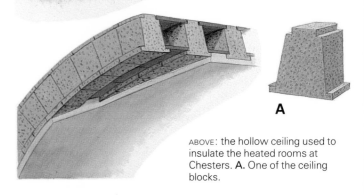

ABOVE: the hollow ceiling used to insulate the heated rooms at Chesters. **A.** One of the ceiling blocks.

LEFT: **1** A ring with three strigils, a mixing dish and an ointment jar.
2 An ointment jar.
3 A strigil used to scrape off sweat and grime. These were all found in the baths at Pompeii in Italy.

ABOVE: two bath house shoes with thick wooden soles found at Vindolanda. Bathers were likely to burn their feet on the hot floors.

BELOW: a reconstruction of the bath house at Vindolanda.
1 Changing room (apodyterium).
2 Cold room (frigidarium) with cold plunge bath. The door on the right leads to the warm room (tepidarium).
3 Hot steam room (caldarium) with hot plunge baths.
4 Boiler room. The furnace heated the hot (**5**) and warm (**6**) tanks as well as the hot and warm rooms.
7 Cold water tank.
8 Hot dry room (laconicum).
9 Latrine.

ABOVE: altar dedicated to Jupiter Optimus Maximus found in the commanding officer's house at Vindolanda.

ABOVE RIGHT: altar dedicated to Fortune found at Vindolanda.
RIGHT: altar dedicated to Silvanus Cocidius from Housesteads.

The final act

The third century was a period of disaster. The empire was torn apart by barbarian invasions and civil wars. The German frontiers suffered terribly but Britain, protected by the Channel, suffered less.

The barbarian invasions forced the Romans to change their strategy; frontier forces were reduced and a mobile army, composed mainly of cavalry, was developed. This army, which was held well back from the frontier, could be rushed to any trouble spot.

The reduction of the frontier forces is well illustrated by the conversion of the barracks at Housesteads into a series of small houses each probably containing one soldier and his family. The garrison, once 800 men, seems to have been reduced to about 100.

It was a period of change. The worship of the traditional Roman gods, often in Celtic guise, gradually waned as eastern religions began to take over. The bull cult of Mithras, in some ways very similar to Christianity, was very popular amongst army officers. Four temples to Mithras have been found on Hadrian's Wall including one at Housesteads. Christianity became the official religion of the empire in the fourth century but it appears to have made little headway in the army.

Peace was restored by the beginning of the fourth century but Britain faced a double threat from the Scottish tribes in the north and sea-borne Saxon raids in the east. In AD 367 the Scottish tribes joined forces with the sea raiders and invaders. Britain was plunged into chaos. The invaders were driven out but yet another civil war began and large numbers of troops were moved to the continent.

Britain was left to defend herself as the Goths and Huns burst upon the western empire. The Roman army never withdrew from Britain. For three hundred years the army had recruited Britons and had become British. Cut off from Rome, with no pay, the soldiers became civilians.

Religion

Roman soldiers worshipped many gods. There were the traditional Roman gods such as Jupiter and Minerva, who were worshipped on official festivals and a host of local gods. The Roman state was very tolerant. Any religion was acceptable as long as it did not have offensive practices, such as human sacrifice, and did not condemn the official religion.

Official gods

The importance of the official gods is illustrated by the large number of altars dedicated to Jupiter Optimus Maximus (I.O.M.), Jupiter the Best and Greatest, which have been found. Altars of this type were set up annually at the edge of the parade ground. Several, dedicated by the First Tungrian Cohort, have been found at Housesteads.

Some units honoured particular gods. Hercules was worshipped by the Tungrians at Housesteads. Mars was worshipped at Vindolanda.

Native gods

Wherever the Roman army went and recruited locally, native gods entered the army along with the new recruits. Often the native god was combined with a Roman god with similar characteristics. The Germanic warrior god Thincsus became Mars Thincsus and was worshipped under this name at Housesteads. The British god Cocidius was also worshipped at Housesteads where he was associated with Silvanus, the Roman god of hunters.

RIGHT: reconstruction of the Mithraeum at Housesteads. The scene shows the ritual feast. The temple was kept in partial darkness to imitate the gloomy cave in which Mithras lived.

Mithras

Mithraism was a very old eastern religion. It was brought to Europe by the Fifteenth Legion when it was transferred from the eastern frontier to the Danube.

Mithraic religion concentrated on the struggle between light and darkness, good and evil. Mithras, lord of light, killed the bull, releasing creative power for mankind. This scene is the centre piece of every Mithraic temple (Mithraeum). It was a mystery religion with severe initiation rites. The cult became very popular amongst officers in the third century AD.

ABOVE: stone carving from Housesteads showing Mithras being born from the rock.

LEFT: one of the Mithraic torch bearers from Housesteads.

The Housesteads Mithraeum

The Mithraeum at Housesteads lies in the valley south of the fort. It conforms to the normal plan for such a temple.

There is an aisle down the middle with a raised platform on either side on which the worshippers reclined to eat the ritual meal. On either side of the aisle were statues of two torch bearers, one with his torch pointed down and the other with his torch pointed up. These appear in the main bull killing scene and are common to all Mithraic temples.

Index

Peter Connolly is an honorary research fellow of
the Institute of Archaeology, University College,
London, and a fellow of the Society of
Antiquaries. He is the author and illustrator of *The
Roman Army, The Greek Armies, Hannibal and the
Enemies of Rome, Pompeii, Greece and Rome at
War, Living in the time of Jesus of Nazareth, The
Cavalryman, The Legionary,* and *The Legend of
Odysseus* for which he won *The Times Educational
Supplement* Information Book Award.

The author wishes to thank Dr Brian Dobson of
Durham University, and Mr Mark Hassall of the
Institute of Archaeology, University College,
London for their advice and help in checking the
manuscript and illustrations.

Oxford University Press, Walton Street, Oxford OX2 6DP

Oxford New York Toronto
Delhi Bombay Calcutta Madras Karachi
Petaling Jaya Singapore Hong Kong Tokyo
Nairobi Dar es Salaam Cape Town

and associated companies in
Berlin Ibadan

Oxford is a trade mark of Oxford University Press

© Peter Connolly 1991

The CIP catalogue record for this book is available from the
British Library.

First published in the United States in 1991
Library of Congress catalog card number 90-53538

ISBN 0 19 917108 4

Phototypeset by Pentacor PLC, High Wycombe, Bucks.
Printed in Hong Kong